Technology Skills for Kids

Good Help Is (Not) Hard to Find

www.technologyskillsforkids.com

Hello and welcome!

This is **book #10** in a series of picture books created to help kids make productive use of their computers and other devices while staying safe online.

Any words or phrases in this picture book series that are in **bold orange** will be explored further in this book's **supplemental materials**.

"Admitting you need help is not a sign of weakness but a sign of strength and self-awareness."

This is a "deleted scene" from the Spanish edition of

Sally Book Bunny and the Search for the Lost Key

– a story in which Sally Book Bunny gets some really good help!

The process for getting good help starts with a clear goal or question.

Sally Book Bunny could not connect the device she needed to use for her presentation to the Internet.

In the midst of technology use, there will often be obstacles and unexpected happenings. (Embrace the challenges!)

A **well-meaning jackass** suddenly appeared!

Well-meaning jackasses tend to **oversimplify** technology-related tasks that require more thought and planning.

Well-meaning jackasses may cheerfully give you the wrong end of the stick.

Well-meaning jackasses will want you to do things **their** way.

The app that the well-meaning jackass giddily championed was a dud. It was kludgy and actually proved to be **harmful** to Sally's goals.

Good help is hard to find...

Seek out help from those who are not only tech-savvy but also exhibit good character.

KNOWLEDGEABLE,
MODEST AND
CARING

Tech Wizard Mike took the time to understand what Sally wanted to do and the difficulties she was having – and **then** proposed a thoughtful solution.

CONNECTED

Sally sought out help from a good egg.

Sally recognized that Tech Wizard Mike was not a bad egg…

…nor was he a silly egg.

And Tech Wizard Mike was certainly not a self-centered, know-it-all egg!

You can **search the web** for any perplexing events that occur during your technology use – using another device if necessary.

An online community or virtual community is a group of people with a shared interest or purpose who use the Internet to communicate with each other.

People in online communities will, for the most part, be **good eggs** who are knowledgeable about a particular subject and simply want to help.

You can search for an **online support community** for the device, operating system or online service that you need help with.

But please Bee Careful as there will **also** be "bad eggs" (e.g. people who want to sell you something you do not need) and "silly eggs" (e.g. people who would never admit they do not know something).

If you know which device you are using (e.g. brand and model), you can **search the web** for its tech support contact information.

Tech Wizard Mike, "Tabby" and "Tough Cookie" are not really support specialists.

Good eggs are knowledgeable but are not looking to project their self-importance. They also have no **technology-related ax to grind**.

A good egg will approach technology-related tasks and conundrums with the same basic methodology they would use to **brew a nice cup of tea**.

"Frosty" got locked out of one of her social media accounts due to her forgetting her username and password.

After getting **good help** and regaining access to her social media account, Frosty was presented with **Tabby's Guide to Thoughtful Tea Drinking** (which is a passwords book cleverly disguised as a cat fancier's tea drinking journal).

A yoga instructor who we will call “Celia” wanted to connect her new smartphone to some much older speakers — she had a clear goal.

We needed to know **which** smartphone Celia wanted to use and **which** speakers she had access to.

www.tabbyfrisk.com
Tabby Frisk
Tabby Frisk
iPhone 14 connection to older Kenwood speakers
Tabby Search
I'm Feeling Lucky

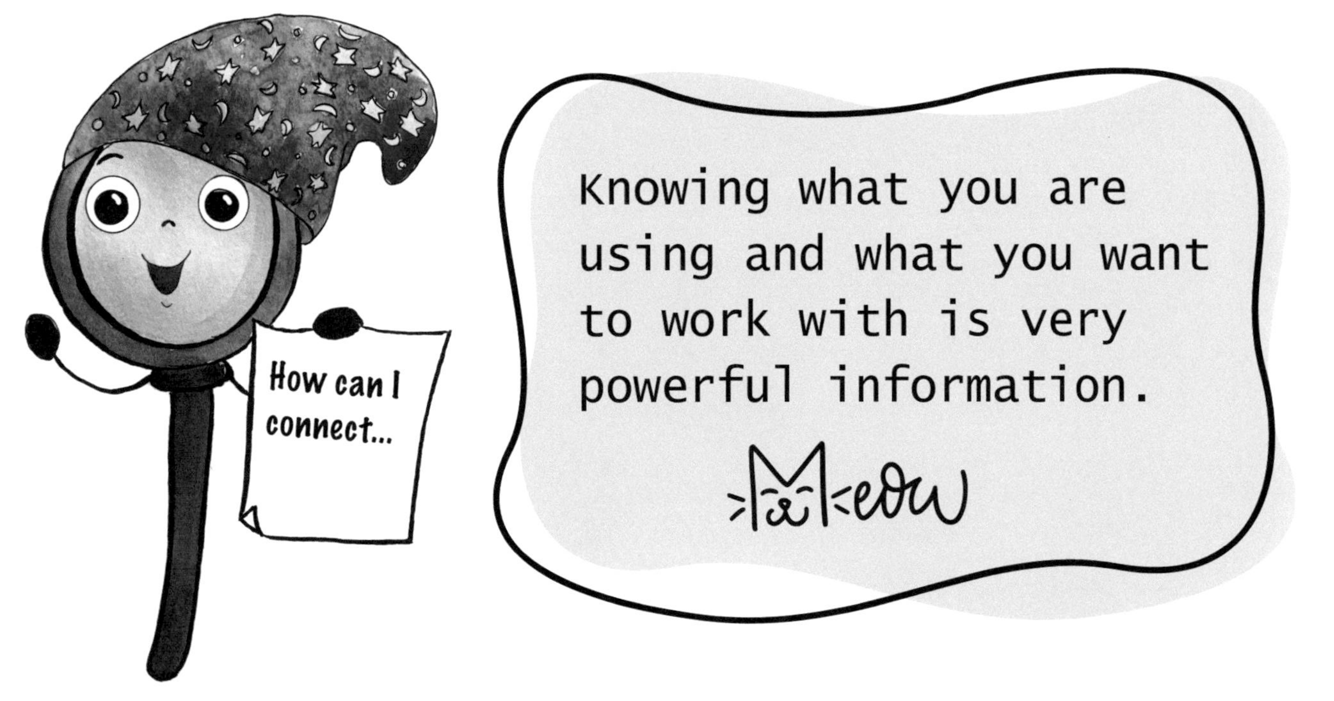
How can I connect...
Knowing what you are using and what you want to work with is very powerful information.
Meow

phew!
CONNECTION MADE!

A clear goal and positive teamwork lead to a successful conclusion.

Shop 'til You Drop, Arthur.

“Arthur” had a clear goal. He had a visualized outcome.

Arthur wanted to shop online and he needed help with the technology-related "nuts & bolts."

A source of good help will help you understand what is needed for any technology-related project.

Tech Wizard Mike broke down Arthur's online shopping project into "action steps" and encouraged him to take good notes.

Tech Wizard Mike carefully taught Arthur how to search for what he needed, how to proceed methodically and how to carefully fill out forms.

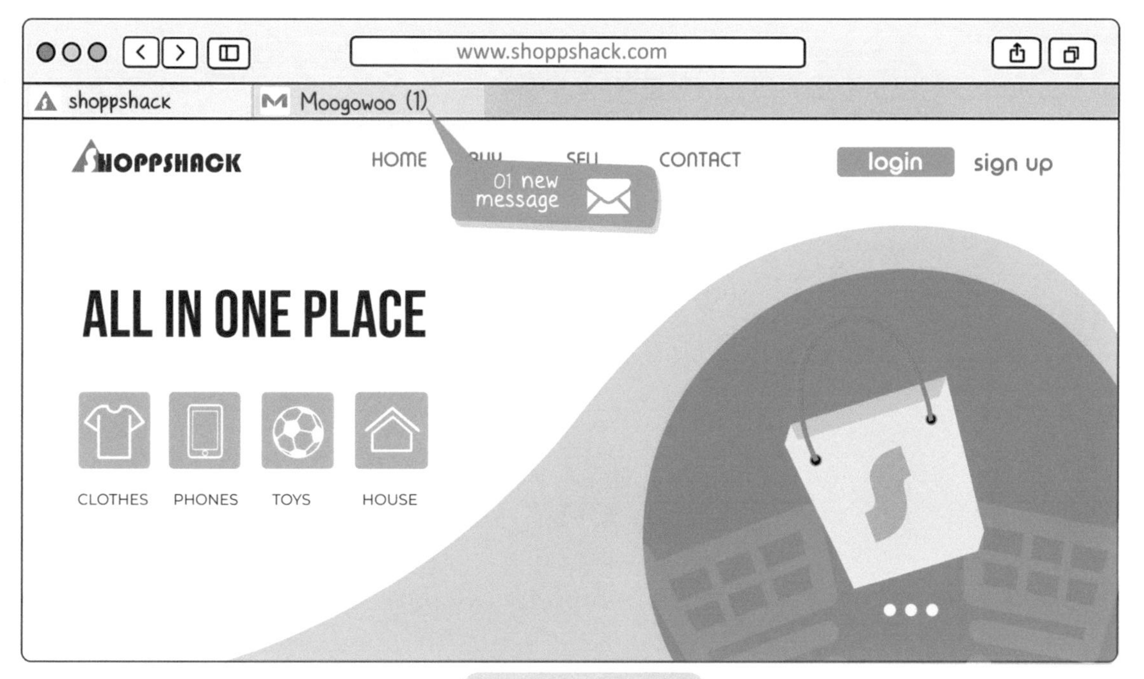

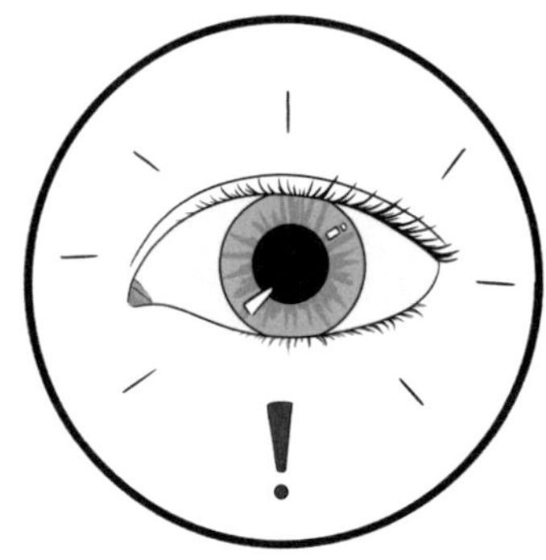

Visual cue!

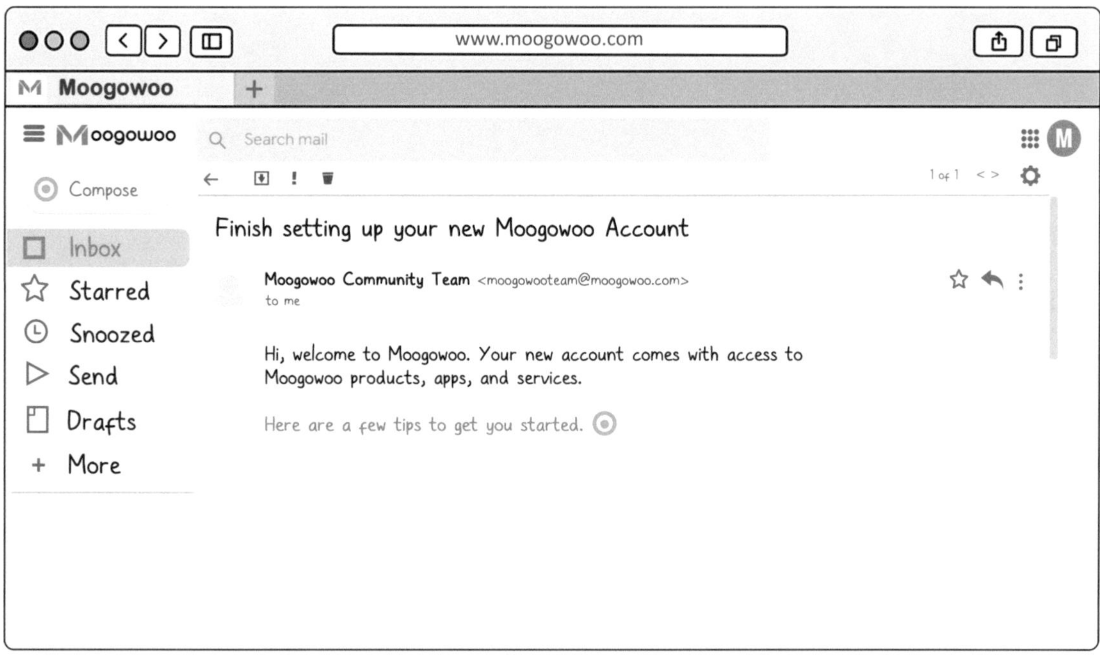

Carefully read and follow **instructions**.

Even a wizard needs to document his technology use.

WEBSITE
MooGoWoo
URL
www.moogowoo.com
USERNAME
arthurdegray32@
moogowoo.com
PASSWORD
cupOftea!_573
NOTES email
address used with
shoppshack
SUGAR
SUGAR

Avoid "Know-It-Alls." They can be identified by their assumed superiority and lack of true **empathy**. They are usually not very thoughtful.

Take everything a know-it-all tells you with at least several grains of salt — and by the way, think twice about having such a person for a friend.

A know-it-all's main goal is to project their **self-importance** and will do so any which way they can.

A source of good help will **not** make things harder than they have to be.

A source of good help will patiently walk you through the problem-solving process.

Getting good help needs to start with a clearly defined issue or a specific question such as:

"Why is my home Internet so slow?"

A Wi-Fi thief is a possible reason for slow Internet and it is also an example of an unexpected happening.

This multitasking barista girl cheerfully gave us the lowdown on how to access the coffee shop's Wi-Fi network.

Coming together is the Beginning

Keeping together is Progress

Working together is Success

1. identify the problem

2. explore information and create ideas

5. evaluate the results

Sally's Problem Solving Loop

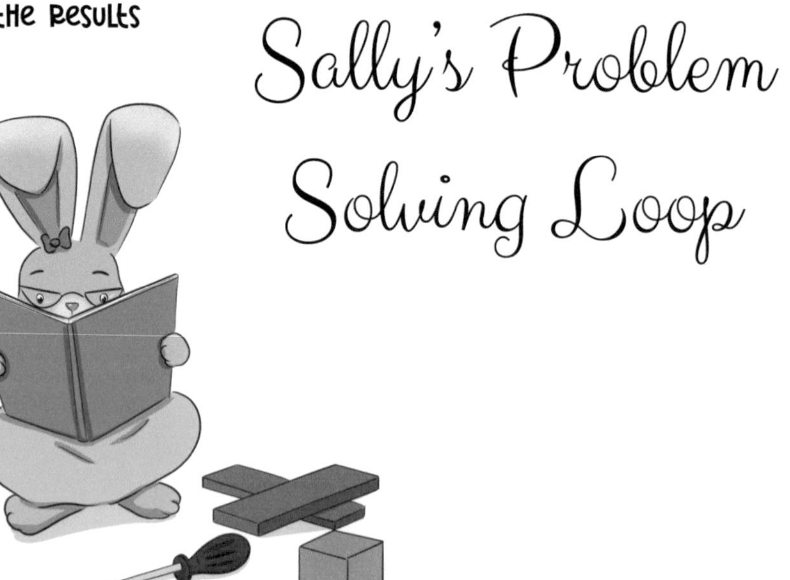

4. build and test the idea

3. select the best idea

A positive collaboration can help you see what you may otherwise have missed.

Name: ______________________ Date:

Quiz for Technology Skills for Kids
Book #10
~ Getting Good Help ~

Read the sentences and circle true (T) or false (F).

1. In the midst of technology use, there will often be obstacles and unexpected happenings. T / F
2. Well-meaning jackasses tend to oversimplify technology-related tasks that require more thought and planning. T / F
3. Ask for help from those who exhibit good character (online and offline). T / F
4. You can search the web for possible solutions to a clearly-defined solutions. T / F
5. New devices (e.g. a smartphone) can never be used with older technologies (e.g. speakers). T / F
6. A recipe for successfully completing a technology-related project is having a clear goal and knowing what you have to work with. T / F
7. A source of good help will teach you how to fish as opposed to handing you a fish. T / F
8. There is never any need to take notes for a clearly defined technology-related project. T / F
9. Seek help from "Know-It-Alls" whenever possible. T / F
10. A well-meaning jackass would never give you a "bum steer". T / F
11. You can use <u>any</u> app for any clearly defined technology-related task. T / F
12. When it comes to technology use, <u>any day</u> can be an "anything can happen day". T / F

Please email techwizardmike@gmail.com for the accompanying quiz and answer key for this book.

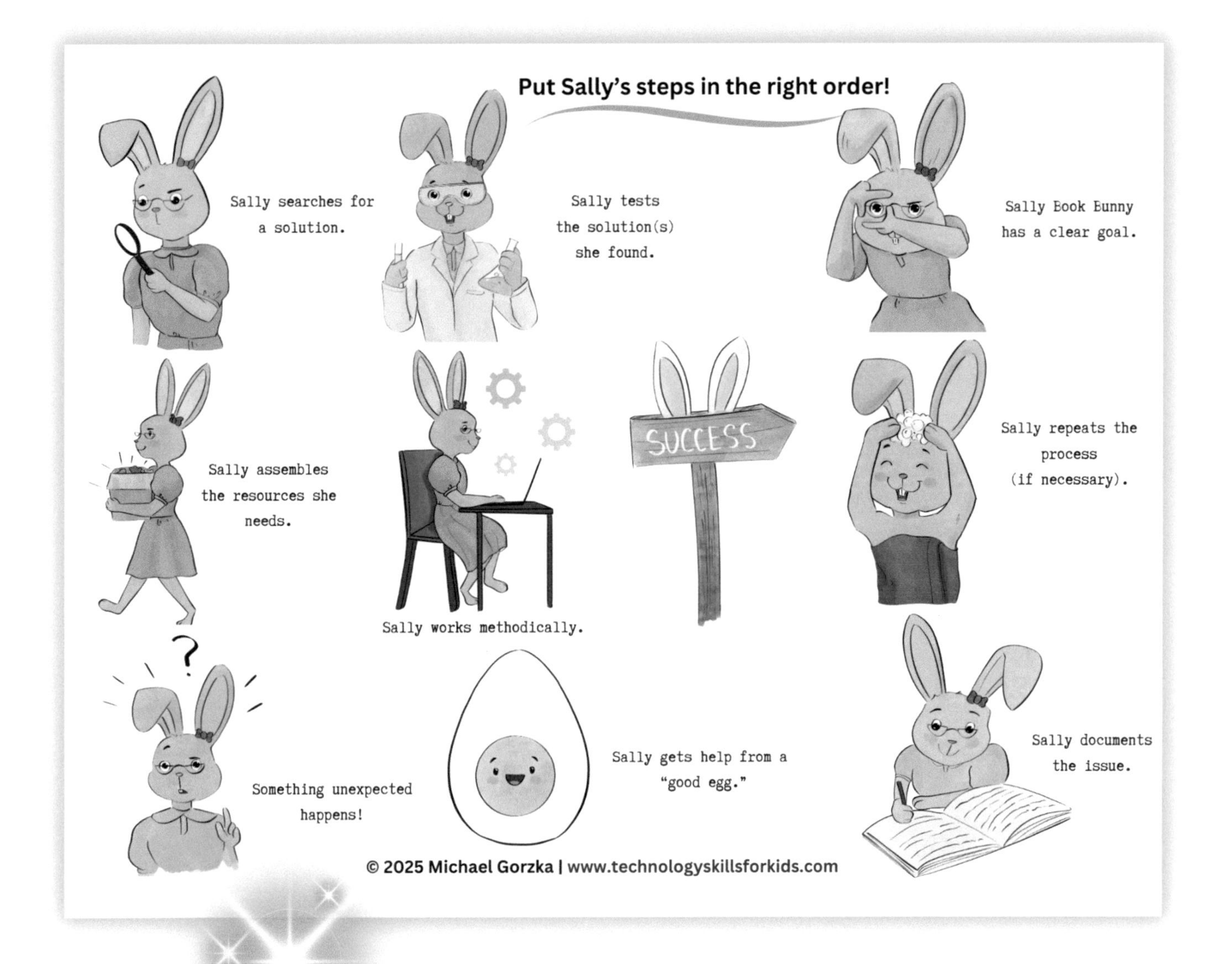

Please email techwizardmike@gmail.com for the accompanying activity and answer key for this book.

Visit www.technologyskillsforkids.com for more tech skills for kids including blog posts, videos and **book #11** in this series which features one of Tech Wizard Mike's favorite tech use tales: "Ed loves Trains."

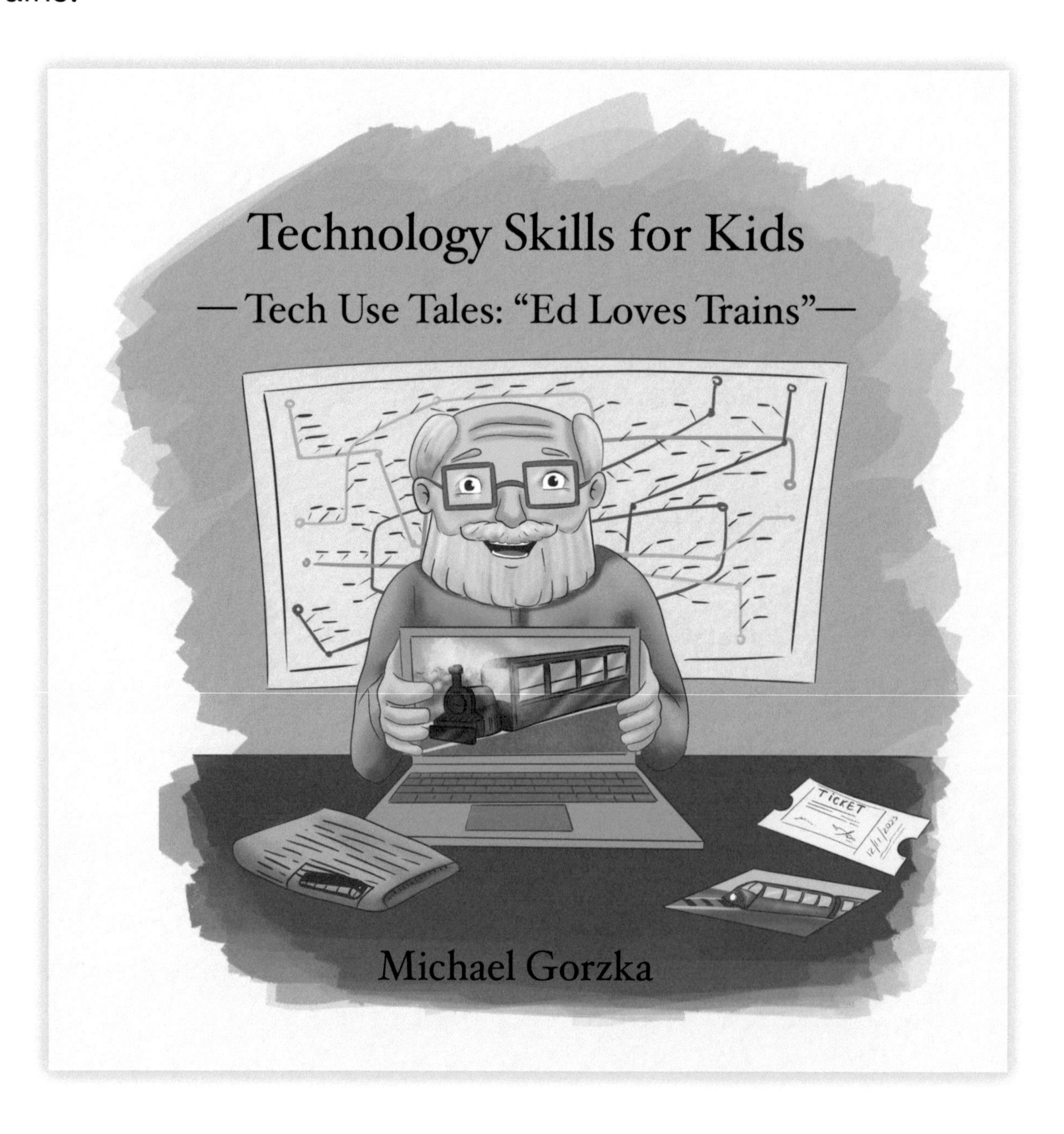